# Objective Reality Is a Myth

**Also by The Quantum Shifter**

*The Truth of Reality*
*The Biggest Misconceptions About Manifestation*
*Why Isn't It Here Yet?*
*Manifest In Steps*
*Identity Shifting*
*One Decision Creates a New Reality*
*Self-Concept Selects Your Reality*
*It's Just a Reflection*
*Manifest Faster*
*Identity Shifting*
*I AM PERFECT*
*Embody a Wealth Identity*
*Mental Diet Mastery*
*Stop Spiraling While Manifesting*
*Emotional Mastery*
*Tell Yourself This When You're Depressed*
*Manifesting Money 101*
*Money From Nothing*
*You Are The Bank*
*Money Mindset For Single Moms*
*Your SP Loves You*
*Manifest A Celebrity*
*Have The Best Sex of Your Life*
*Lose Weight The Mental Way*
*Manifest Away "ADHD"*
*The Writer Everyone Reads*
*You Don't Need 3D Proof*
*Action Is Optional*

*The Positive Delusion*
*Cause and Effect Is a Lie*
*Objective Reality Is a Myth*
*Revision*
*Collapse & Command*

# Objective Reality Is a Myth

**How to Stop Believing in "Facts" and Start Creating Your Own Reality**

**The Quantum Shifter**

Identity Press

Connect with The Quantum Shifter

Website: www.TheQuantumShifter.com

Facebook: www.Facebook.com/TheQuantumShifter

ISBN: 978-1-972432-26-6

Published by Identity Press



**These works were written over many years during the author's ongoing study of identity and reality creation.**

**They are now being published for the first time under the pen name *The Quantum Shifter*.**

YOUR REALITY IS WHAT YOU SAY IT IS.

# Contents


Before You Continue XIII

Introduction: The Myth of Objectivity 1

1. The End of Objectivity 5
2. Why Evidence Isn't Universal 11
3. The Quantum View 17
4. Everyone's Right (But Only in Their World) 25
5. The Mirror Principle 31
6. Collapsing Collective Truths 37
7. Freedom From Proof 43
8. Identity as the Only Law 49
9. Choosing Your World Daily 55

Conclusion: The Power of Subjective Sovereignty 61

About The Author 65

## *Before You Continue*

This book will make the most sense if you are open to a simple idea: Your reality is not fixed. It is not happening to you. It is reflecting what you assume is true.

If you believe that reality is solid, unchangeable, and determined by circumstances, parts of this book may feel confusing or difficult to accept. And that's okay. You don't need to force yourself to believe anything. You can simply set this book down and return to it later, if and when it feels right.

But if something in you already senses that reality may not be what it seems—if you've noticed patterns that don't quite make logical sense—or if you've experienced moments where things shifted in ways you couldn't explain— then this book will likely feel familiar.

Not because it's teaching you something new, but because it's putting words to something you already know. Read it slowly. Let it land. And most importantly: Don't try to "do" anything. Just see what you notice.

What you call "facts" are just reflections of what you already assume.

# Introduction: The Myth of Objectivity

Most people live as if reality is fixed, shared, and objective. They believe there's one world out there, and we're all just moving through it. They believe facts are facts, evidence is evidence, and what "everyone sees" must be true.

But that belief is the biggest illusion of all.

**There is no objective reality.**

What looks like solid, unquestionable truth is always filtered by the story you hold. Your assumptions shape what you notice, what you interpret, and what you experience. What you call "objective" is just repeated agreement. **What you call "facts" are just reflections of what you already assume.**

## Why We Crave Objectivity

People cling to objectivity because it feels safe. If reality is shared and solid, then they don't have to take responsibility for creating it. They can point to circumstances, evidence, and causes as if those forces were stronger than their own word.

But look closer, and you'll see cracks in the story.

- Two people can experience the same event and tell completely different stories about what happened.
- Two scientists can run the same experiment and interpret opposite conclusions.
- Two friends can read the same headline and live in completely different worlds afterward.

If reality were truly objective, how could this happen?

## The Subjective Truth

Reality bends to identity. It bends to assumption. It bends to you.

What you believe about money, love, health, and success doesn't just color your perception — it creates the stage you live on. Someone else, living under a different set of assumptions, experiences a different "objective reality." But they're not in yours, and you're not in theirs.

There is no shared world. There are only reflections of story.

## The Pivot This Book Offers

This book is not about convincing you that objectivity is false. It's about showing you the freedom that comes when you drop the myth.

Once you stop trying to live by "common reality," you gain the authority to declare your own. You stop debating, waiting, and proving. You simply say what's true for you, and let reality obey.

The end of objectivity is not chaos. It's sovereignty. It's the clearest, simplest way to live.

"There is no objective reality — only story."

## Chapter 1

# The End of Objectivity

The world teaches you to believe in a single, shared reality. Science, culture, religion, and education all reinforce the idea that there are universal truths that apply to everyone, regardless of perspective.

But when you look closer, that idea doesn't hold up. What we call "objective reality" falls apart the moment you realize: every experience is filtered through story.

### Two People, One Event — Two Realities

Imagine this: two people walk through the same rainstorm.

- Person A grumbles, "*This is miserable. My day is ruined.*"
- Person B smiles, "*The rain feels refreshing. I love this weather.*"

Same storm. Two completely different realities.

If the storm were truly objective, it would carry one meaning for everyone. But it doesn't. The meaning only exists because of the story each person applies.

## Science Isn't Exempt

Even science, often considered the gold standard of objectivity, proves the opposite. Experiments don't reveal one "truth." They reveal what the observer is set up to find. Quantum physics has already shown this: the act of observation changes the result.

The conclusion? Nothing exists without perspective. There is no reality independent of the one observing it.

## Why "Facts" Differ

Think of debates about politics, money, or health. Each side presents "facts." Each side is convinced they're right. And within their reality, they are right.

Facts aren't universal. They're reflections. The same event, statistic, or headline can mean abundance for one person and disaster for another. What's objective about that?

## The Cracks in Consensus

Consensus is not objectivity. It's just a lot of people agreeing on the same story.

When enough people repeat an assumption, it gets labeled as fact. But the moment you step out of agreement, you step into a different reflection.

That's why you don't have to live by collective rules. You don't need the majority to agree with you. You only need to stabilize your own story.

## Identity Statements

- "There is no objective reality — only story."
- "Consensus is not truth. It's agreement."
- "My observation creates my world."
- "What others call fact, I call reflection."

## Exercise: Spot the Subjective

1. Think of a recent event where two people interpreted it differently. Example: a layoff — one saw disaster, another saw freedom.

2. Write down the two interpretations.

3. Notice: the event itself didn't dictate meaning. The story did.

4. Declare: "*I choose my story, and my story creates my world.*"

## Closing Reminder

The myth of objectivity keeps people trapped in stories they didn't choose.

The moment you drop the myth, you see the truth: there is no one world. There is only your world, created by your story.

# Integration Notes

"My story creates the proof I see."

# Chapter 2
# Why Evidence Isn't Universal

People cling to evidence because it feels solid. "Show me the proof," they say, as if facts exist outside of interpretation. But evidence is never neutral. It always bends to the assumptions you bring into it.

## How Belief Shapes Proof

If you assume money is hard to get, you'll notice every struggle, every bill, every rejection. That becomes your "proof."

If you assume money flows easily, you'll notice every discount, every refund, every unexpected payment. That becomes your "proof."

Both people can live in the same world, yet both collect completely different evidence. Which one is right? Both. In their own story.

## Evidence as a Mirror

Reality doesn't hand you neutral facts. It mirrors your expectations.

- Believe you're unlucky? You'll spot coincidences that prove it.
- Believe you're magnetic? You'll spot synchronicities that confirm it.
- Believe relationships are hard? You'll find proof of struggle.
- Believe love is natural? You'll find proof of ease.

Evidence isn't the cause. It's the echo.

## Why Proof Keeps You Stuck

Chasing proof is the fastest way to stay in the old story.

- You say, "*I'll believe I'm wealthy once I see the money.*"
- Reality mirrors back, "*You don't see it yet.*"

The cycle never ends, because you're waiting for reflection instead of declaring the cause.

## Shared Evidence, Different Realities

Even when two people look at the same data, they interpret it differently.

- One investor sees a falling market as disaster.
- Another sees the same market as opportunity.

Same evidence. Two opposite realities. Each feels true because each is filtered through story.

## Identity Statements

- "Evidence is never objective — it always bends to story."
- "Proof is not cause. It is reflection."
- "I don't wait for evidence. I declare truth first."
- "My story creates the proof I see."

## Exercise: Flip the Proof

1. Write down one area where you've been waiting for evidence. Example: "*I'll believe money flows once I see more in my account.*"
2. Cross it out.
3. Replace with: "*I declare money flows easily. My account reflects it now.*"
4. Notice how quickly small reflections begin to show up.

## Closing Reminder

Evidence isn't universal. It isn't neutral. It's a mirror of what you assume.

**When you stop chasing proof and start declaring truth, reality has no choice but to give you new evidence.**

# Integration Notes

"Nothing exists until I assume it does."

# Chapter 3
# THE QUANTUM VIEW

If you've ever wondered why your world feels so personal – so tailored to your own thoughts, fears, and expectations – quantum physics offers a clue.

At its core, quantum science shows that nothing exists as a solid fact until it's observed. Until attention is placed on it, reality is a field of possibilities. The moment you observe, it collapses into one outcome.

This isn't far from what you've already been living: your assumptions act as observation. They decide which possibility solidifies into your reality.

## Nothing Exists Until You See It

Think of walking into a room. Before you enter, you don't know what's inside. It exists only as potential. The moment you open the door, the details "snap" into place — the lighting, the furniture, the mood.

Your mind doesn't passively record. It actively chooses.

## Assumption as Observation

In quantum mechanics, the observer isn't separate from the experiment — the observer influences the outcome.

In reality creation, assumption works the same way.

- Assume struggle, and you collapse timelines where money is scarce.
- Assume wealth, and you collapse timelines where money flows easily.
- Assume rejection, and you collapse timelines where love feels difficult.
- Assume magnetism, and you collapse timelines where love finds you.

Observation = assumption. Reflection = reality.

## The Illusion of Objectivity

Science once believed it could describe an independent, objective world. Quantum discoveries shattered that illusion. Reality is not independent of the observer — it exists through interaction.

The same is true in your daily life. What you call "objective reality" is only the reflection of what you've observed through your assumptions.

## Everyday Example

- Two people walk into a networking event.
    - One assumes, "*Nobody likes me.*" Their reality reflects coldness and awkward conversations.
    - The other assumes, "*I'm magnetic.*" Their reality reflects warmth, attention, and opportunities.

Same event. Different observations. Different collapsed realities.

## Identity Statements

- "Observation is creation."
- "Nothing exists until I assume it does."
- "I don't discover reality – I collapse it."
- "The world doesn't show me truth. It shows me me."

## Exercise: Practice the Collapse

1. Think of a situation you're about to step into (a meeting, a date, a purchase).
2. Write down the story you want to see. Example: "*I am magnetic. People light up around me.*"
3. As you step in, hold that assumption.
4. Notice how the scene unfolds around the version you declared.

## Closing Reminder

The quantum view shows what you already know: nothing is fixed. Nothing is independent. Nothing is objective.

Reality waits for your observation. And your observation is the story you choose.

# Integration Notes

There are only
stabilized stories,
mirrored back to those
who believe them.

Chapter 4

# Everyone's Right (But Only in Their World)

If you've ever argued with someone – about politics, religion, money, health, or even sports – you've probably noticed something: no matter how much "evidence" you present, they don't change their mind. And no matter how much evidence they throw back, you don't change yours.

That's because both of you are right. But only in your own world.

## Why Arguments Never Resolve

Arguments drag on because both people are fighting for their reflection. Each one has stabilized a story, and reality keeps feeding them proof.

- To someone who assumes money is scarce, every headline confirms it.
- To someone who assumes money flows

easily, every event confirms that too.

Neither is lying. Neither is delusional. Both are right — within their chosen story.

## Multiple Realities, Side by Side

It's easy to believe there's just one "shared" world, but in practice, each person is living in a personal projection.

- One person lives in a world where hard work pays off.
- Another lives in a world where wealth comes with ease.
- One sees relationships as constant struggle.
- Another sees them as natural and harmonious.

These realities can exist side by side, like two movies playing in the same theater, but you only see the one you've bought a ticket for.

## Why This Brings Peace

When you realize everyone's right in their own world, you stop wasting energy trying to convince them. You no longer need agreement to validate your story.

You can let them keep their reflection while you keep yours.

### Everyday Example

Think about fitness.

- Person A believes: "*Losing weight is impossible after* 40." Their world shows them proof: slow results, frustration, struggle.
- Person B believes: "*I'm always fit and healthy.*" Their world shows them proof: energy, compliments, visible results.

Same age, same bodies, different realities. Each is right, because each is living their own story.

## Identity Statements

- "Everyone is right, but only in their world."
- "I don't need agreement – I only need my story."
- "Arguments dissolve when I remember

reflections differ."

- "I stabilize my world, and others stabilize theirs."

## Exercise: Release the Debate

1. Think of one area where you've been arguing with someone (money, politics, relationships).
2. Write down their belief and your belief.
3. Notice how both are right in their own reflections.
4. Say: "*I release the need to convince. My reality is mine.*"

## Closing Reminder

There is no one objective truth for everyone. **There are only stabilized stories, mirrored back to those who believe them.**

Everyone is right — but only in their world.

# Integration Notes

It's all a mirror of your own stabilized assumptions.

Chapter 5

# The Mirror Principle

One of the most freeing — and confronting — truths about reality is this: no one is separate from you. Every person you meet, every reaction you see, every outcome that involves another human being — **it's all a mirror of your own stabilized assumptions.**

This isn't about blame. It's about reflection. Just as a mirror doesn't create your face but only reflects it, the people in your world don't create your story. They reflect the story you've stabilized.

## Why Others Can't Be Separate

If reality is shaped by identity, then everyone you encounter must appear in alignment with who you believe you are. They can't show up outside of your world's script.

- If you believe, "I *am respected*," people mirror respect.

- If you believe, "*I am overlooked,*" people mirror dismissal.
- If you believe, "*I am magnetic,*" people mirror attraction.

Their behavior feels personal, but it's really projection.

## The Trap of Blame

When someone mistreats you, the natural response is to blame them. But blame assumes separation: "*They acted independently of me.*"

In truth, they're acting within the role your assumptions made available. Change the assumption, and their reflection shifts — or they disappear from your story altogether.

## Relationships as Proof

Relationships show the mirror principle more clearly than anything else.

- Person A assumes, "*I'm always betrayed.*" Every partner eventually cheats or lies.
- Person B assumes, "*I'm always chosen.*" Every partner shows loyalty and devotion.

The partners may differ, but the reflection is consistent — because the story behind it is consistent.

## How to Use the Mirror

Instead of resenting the mirror, use it. Every reaction, every interaction, is feedback. It shows you what you've stabilized. If you don't like the reflection, you don't argue with the mirror — you change the identity it reflects.

## Identity Statements

- "No one is separate from me — everyone reflects my story."
- "People can only show me who I assume I am."
- "If I want a new reflection, I choose a new identity."
- "The mirror is feedback, not truth."

## Exercise: Spot the Mirror

1. Think of one recurring pattern in relationships (romantic, family, business).
2. Ask: what assumption about myself is being reflected? Example: "*People don't take me seriously.*"
3. Write a new assumption that declares the opposite. Example: "*I am always valued and respected.*"
4. Watch as interactions begin to mirror the new identity.

## Closing Reminder

People don't act outside of your story. They mirror it.

And once you stop fighting the mirror and start rewriting the identity, relationships become proof of your chosen reality — not prisons of your old one.

# INTEGRATION NOTES

"Only my story is binding in my world."

"Only my story is binding in my world."

# Integration Notes

# Chapter 6
# COLLAPSING COLLECTIVE TRUTHS

The world is full of so-called "truths" that everyone is expected to agree on. Headlines, cultural sayings, and expert opinions all carry the weight of consensus. But consensus isn't law. It's just a collection of repeated assumptions.

When you realize this, you stop being trapped by collective truths. You see them for what they are: optional stories, offered to you like dishes on a buffet. You can say yes. You can say no. You can choose differently.

## News as a Buffet

The news doesn't deliver facts. It delivers stories.

- One headline says the economy is collapsing.
- Another says opportunities are boom-

ing.

- One channel says disaster is imminent.
- Another says growth is unstoppable.

Which one is true? Whichever one you decide belongs to your reality.

## Culture as Programming

Phrases like "*money doesn't grow on trees,*" "*relationships are hard,*" or "*you can't have it all*" aren't wisdom. They're programming. Passed down, repeated, and reinforced until they feel like law.

But they're not laws. They're just collective loops. And you don't have to keep playing them.

## Collapsing What You Don't Choose

**The simplest way to collapse a collective truth is to say: "*That's not my reality.*"**

You don't argue. You don't debate. You don't fight for agreement. You simply decline the dish and choose another.

## Everyday Example

Two people read the same headline about the housing market.

- Person A believes: "I'll *never afford a home*." They see the headline as confirmation and feel defeated.
- Person B believes: "I *always find what I want*." They see the headline as irrelevant — and shortly after, the perfect home appears for them.

The headline wasn't the cause. The story was.

## Identity Statements

- "Collective truths are just optional stories."
- "Consensus is not law."
- "I collapse what I don't choose."
- "Only my story is binding in my world."

## Exercise: Collapse a Collective Story

1. Write down one cultural or news story you've been agreeing with. Example:

*"The economy is bad, so money is scarce."*

2. Cross it out.

3. Replace it with your own declaration. Example: "*The economy doesn't control me. Money flows to me endlessly.*"

4. Each time the collective story appears, remind yourself: "*That's not my reality.*"

## Closing Reminder

The collective can shout as loudly as it wants, but it doesn't decide your world.

Consensus is not truth. Your story is.

# Integration Notes

"Reality follows me,
not the other way
around."

Chapter 7

# Freedom From Proof

One of the biggest traps people fall into is believing they need proof before they can believe. *"I'll feel wealthy once the money shows up." "I'll feel loved once someone chooses me." "I'll feel successful once I get the results."*

But waiting for proof is what keeps you stuck. Proof is always the reflection, never the cause.

## Why We Crave Proof

Proof feels comforting because it takes the pressure off. If reality is objective, then proof means "I'm safe to believe now."

But what actually happens is the opposite. **By waiting for evidence, you keep declaring, "I *don't have it yet*."** Reality mirrors that back, and the cycle continues.

## Proof Is Always Late

The 3D is never the first mover. It only reflects what you've already stabilized.

- Declare, "I *am wealthy.*" Proof shows up after.
- Declare, "I *am chosen.*" Proof shows up after.
- Declare, "I *am successful.*" Proof shows up after.

If you keep waiting for proof first, you'll keep waiting forever.

## The Trap of External Validation

When you need others to agree before you believe, you hand them authority over your world. You say, "*I'll stabilize my story once they approve.*" But their reflection is still your story — which means you've been giving away power you never actually lost.

## Living Without Proof

The moment you drop the need for proof, you feel free. You no longer argue with reality. You no longer wait for consensus. You no longer delay your declaration.

You simply decide.

And reality, as always, follows.

### Everyday Example

- Old pattern: "*I'll believe I'm successful once I hit six figures.*"
- New story: "I *am successful now. Six figures reflects that.*"

The success wasn't caused by the number. The number was the mirror.

## Identity Statements

- "Proof is never the cause. It's always the echo."
- "I don't wait for evidence. My word is law."
- "I live from declaration, not validation."
- "Reality follows me, not the other way around."

## Exercise: Drop the Proof

1. Write down one area where you've been waiting for evidence. Example: "*I'll feel secure once I pay off debt.*"

2. Cross it out.

3. Replace with: "*I am secure now. My world reflects it.*"

4. Each time you look for proof, remind yourself: "*I don't wait. I declare.*"

## Closing Reminder

Proof isn't freedom. It's delay.

The freedom comes when you drop the need for evidence and live from your declaration now.

# Integration Notes

"My I am decides my world."

Chapter 8

# Identity as the Only Law

Every assumption, every reflection, every story comes down to one root truth: **identity rules reality.**

"I am" is the strongest declaration you can make. Whatever follows those words becomes the law of your world. Not because it's "objectively true," but because reality has no choice but to mirror it.

## Why Identity Always Wins

You can try affirmations, vision boards, or rituals. But if your identity contradicts them, the identity will win.

- If you say, "I *attract wealth*," but deep down your identity is "I *never get ahead*," the identity wins.
- If you say, "I *want love*," but your identity is "I'*m always rejected*," the identity wins.

- If you say, "*I'll be successful someday*," but your identity is "*I'm struggling*," the identity wins.

Identity overrides effort, tools, and even desire.

## "I Am" as Law

The moment you declare "I am," you've chosen your reflection.

- "*I am wealthy*." Reality bends to show it.
- "*I am chosen*." People mirror it back.
- "*I am successful*." Opportunities line up.
- "*I am stable*." Life feels calm and supportive.

"I am" doesn't describe. It commands.

## The Illusion of Facts

So-called "facts" are only stable because identity makes them so.

- Someone who says, "*I'm always broke*," will always find facts to support it.
- Someone who says, "*I'm always abundant*," will always find facts to support

that.

Facts don't create identity. Identity creates facts.

### Everyday Example

Two people apply for the same job.

- Applicant A: "*I'm never chosen.*" Reflection: rejection.
- Applicant B: "*I'm always chosen.*" Reflection: offer.

Same job. Same interview process. Different identities, different realities.

## Identity Statements

- "Identity is the only law."
- "My I am decides my world."
- "Facts are reflections, not causes."
- "I don't wait for proof — I declare who I am."

## Exercise: Choose Three "I Am" Statements

1. Write down three declarations of identity you want to stabilize. Example: "*I am wealthy. I am loved. I am successful.*"

2. Repeat them daily — in the mirror, before bed, while doing daily routines.

3. Each time an "old fact" contradicts them, remind yourself: "*Facts bend to identity. This reflection is temporary.*"

4. Stay consistent until the reflection stabilizes into your new normal.

## Closing Reminder

There are no laws outside of identity. There are no facts stronger than "I am."

Once you stabilize identity, reality bends. Always.

# Integration Notes

"I say yes or no, and
reality obeys."

Chapter 9

# CHOOSING YOUR WORLD DAILY

If there is no objective reality, then every day becomes a practice of choice. Stories, headlines, and conversations present themselves constantly — like a buffet. You don't have to eat everything on the table. You choose what goes on your plate.

This is where reality creation becomes simple. You don't argue. You don't overthink. You don't wait for proof. You simply say: "*Yes, that's mine*" or "No, *not in my reality*."

## The Buffet of Stories

Life presents you with endless options:

- A news article says the economy is crashing.
- A friend complains about relationships always failing.
- A family member says money only

comes through hard work.

Each one is a dish at the buffet. Do you have to take it? No. Do you have to debate it? No. You simply decline.

## The Power of "Yes" and "No"

Reality collapses instantly around what you agree to.

- See a story of abundance → say "Yes, *that's mine.*"
- Hear a story of struggle → say "No, *not in my world.*"

The moment you respond, you've made your choice. Reality doesn't force-feed you. It obeys your acceptance or rejection.

## How to Practice

Throughout the day, notice what stories present themselves. With each one, ask:

- Do I want this reflected in my reality?

If yes → declare it yours. If no → collapse it instantly by saying, "Not *my story.*"

## Everyday Example

Scrolling social media, you see two posts:

- Post A: "*The world is falling apart.*"
- Post B: "*Opportunities are everywhere.*"

You choose B. You stabilize the reflection of opportunity. The choice was yours all along.

## Why This Matters

Many people live passively, letting stories attach to them without question. That's how they get trapped in collective loops.

Once you practice daily choice, you realize you are never at the mercy of what you see. You are always at the helm.

## Identity Statements

- "Every story is optional."
- "I say yes or no, and reality obeys."
- "I don't argue with stories. I choose."
- "My daily choices stabilize my world."

## Exercise: Buffet Practice

1. For one day, consciously practice yes/no with every story you encounter.

2. Write down a few examples at night. Example: *Said no to: 'Money is hard.' Said yes to: 'Opportunities flow.'*

3. Notice how much lighter and more in control you feel.

4. Repeat daily until it becomes natural.

## Closing Reminder

Reality doesn't force itself on you. It offers itself.

You're the one who chooses your plate.

# Integration Notes

There is only the story
you stabilize.

# Conclusion: The Power of Subjective Sovereignty

For most of your life, you were told reality was fixed. That facts were facts. That the world was objective, shared, and beyond your control.

But now you see the truth: there is no objective reality. **There is only the story you stabilize.**

## The Weight Lifts

When you stop chasing consensus, proof, and collective truths, life gets lighter. You no longer need the world to agree with you. You no longer waste time arguing over what's "real." You simply declare, and your reality obeys.

## Simplicity Returns

Instead of asking: "*What's true?*" You ask: "*What do I want to be true in my world?*"

Instead of debating: "*Whose facts are right?*" You decide: "*I don't need their facts. I have mine.*"

Instead of waiting: "*When will my life change?*" You live: "*It already has, because I said so.*"

## Sovereignty in Action

Freedom doesn't come from proving your perspective. It comes from remembering you are the source.

- Your identity creates facts.
- Your observation collapses reality.
- Your story is the only law in your world.

That's sovereignty. And it's the opposite of being trapped in "objective reality."

## Final Word

There is no shared world you must conform to. There is only your world – authored by your assumptions, reinforced by your choices, and reflected in your experience.

The myth of objectivity ends here. The freedom of sovereignty begins now.

And the only law that matters is the one you declare.

# About The Author

**The Quantum Shifter** is the pen name of an Ivy-trained psychologist, researcher, and lecturer whose work explores the intersection of identity, consciousness, and reality creation. After completing advanced studies in psychology and spending years teaching and writing at the university level, she shifted her focus toward the deeper question that shaped her life's work:

**How does identity create reality — and why does changing the story of who you are change the world you live in?**

Her publications draw on decades of private research, academic inquiry, and experiential observation. Known for her clarity, precision, and deeply grounded approach, she writes about identity physics, timeline selection, self-concept, and the mechanics of reality in a way that is both accessible and transformative.

Although her academic background is traditional, her pen name reflects the evolution of her work — a nod to the moment she realized that consciousness itself is the most powerful instrument of change. The name stuck, and readers around the world now associate *The Quantum Shifter* with high-level insight, elegant simplicity, and rapid identity transformation.

She currently lives a quiet, contemplative life dedicated to writing, research, and teaching the principles of identity-based reality creation. Her books serve as distilled explorations of the concepts she has spent decades studying — written for readers who are ready to shift who they are, not just what they want.

www.ingramcontent.com/pod-product-compliance
Lightning Source LLC
Chambersburg PA
CBHW031333230926
51848CB00001B/213
*9781972432266*